ISLANDS IN THE SKY

SCENES FROM THE COLORADO PLATEAU

A WISH YOU WERE HERE BOOK

BY
JEFF NICHOLAS
AND
LYNN WILSON & JIM WILSON

SIERRA PRESS, INC.

ISBN O-939365-16-2 (Paper)
O-939365-19-7 (Cloth)

Printed in Singapore
Second Edition

SIERRA PRESS, INC.

P.O. BOX 430, EL PORTAL, CA. 95318

ACKNOWLEDGEMENTS

We wish to thank Ellis Richard of Grand Canyon, Tim Manns of Zion, Edd Franz of Bryce Canyon, Vince Santucci of Petrified Forest and Janet Lowe of Canyonlands N.H.A. for their assistance with this project. We also wish to thank the employees, both past and present, of each Park and Natural History Association on the Colorado Plateau. It is due to their foresight and hard work that wonderlands such as these are still here for each of us to enjoy. It is up to each of us, as individuals, to make certain our own use is consistent with the long-term needs of these natural temples.

DEDICATION

This book is dedicated
to those who stop;

to see, to hear, to smell, to taste, to feel,

not just to know,
but to understand.

CONTENTS

INTRODUCTION 4
PREFACE 4
EARTH 6
ROCK 34
WATER 48
SKY 62
NOTES on the PLATEAU 90

INTRODUCTION

These are your places. They have no name, at least not one I will mention. They are the secrets of the earth. Known but not publicized. Whispered by those who take time to explore them. Found anywhere people cannot easily reach.

This is our earth. Its beauties are for everyone. The right to see, experience, smell, touch and enjoy awaits those with the desire and willingness to walk.

This is the land of the ancient ones. People who melded with Mother Earth and became as one with her rhythms. Far removed from civilization, few have experienced its wildness and wonder. It is a land beyond the scope of our imagination, part Venus, part Mars and part Moon. A little bit of heaven and a little bit of hell.

Land protected, because we are not its protector. Saved because we have not discovered it needs saving. A land where litter does not yet exist and the rare footprint left in the sand is erased by the afternoon breeze.

Take my mind's hand and tour these incredible lands without roads. See visions of a thousand millennia honed to these few photographs. Know that these are wild places still filled with wonder, unaffected and completely untamed by man.

SLOT CANYON, LOCATION UNDISCLOSED

PREFACE

The Colorado Plateau. To those who have been touched by this land, nothing more need be said. But how can you describe this region to someone who has never visited it? The use of superlatives does not suffice, for above all this is a land of subtle beauty. A beauty that slowly, gradually seduces the perceptive visitor. Known variously as The Plateau, Four Corners, Standing Up Country, Red Rock Land or Canyon Country, this is essentially an unknown land, a land that invites, indeed demands, personal involvement and exploration.

seen from the edge of a plateau or mesa, as at Muley, Rainbow, Grandview or Dead Horse Points, the horizon may easily be one hundred miles distant, broken only by the jagged, granitic accoliths of the La Sal, Abajo or Henry Mountains. The landscape between your vantage point and those peaks, however, is a nightmare of 1000-foot canyons, of twisted, eroded, barren slickrock that does not know the meaning of 'straight line'. Viewed from the South Rim of Grand Canyon, North Rim is but ten miles away, yet it would require a seasoned hiker at least two days or a diligent driver more than 200 miles and five hours of travel to get there.

This was the last region in the contiguous states to be explored and mapped. And while it is always wise to carry a map when hiking, it will do the hiker little good in this region. The labyrinthine canyons make concepts like North and South all but meaningless; ultimately it is a land of upstream or downstream.

Seen at river level, this land of hundred-mile horizons is reduced to what lies between here and the next bend in the river. Enclosed by cliffs that tower thousands of feet overhead, a great flat landscape of plateaus and mesas has become a land of cliff and pinnacle, a skyline broken and shattered, one thin slice of cerulean blue sky that is just beyond reach. Following a summer thundershower, the deafening silence of the mesa top is replaced by the thundering roar of rapids casually rolling house-sized boulders along the river bottom.

This twisted, uplifted, fault-blocked, anticlined sun-baked, wind-blasted landscape of buttes canyons, pinnacles, arches, mesas, spires and fin. is colored a palette of red, yellow and brown. Bu we are not talking about crayon colors. We are talking vermilion, white, grey and pink; chocolate, caramel, coffee and butterscotch; burnt sienna, umber, fawn and tan; lavender, copper bronze and gold. A shimmering watercolor tha changes tint and tone with every passing hour o the day, with every passing cloud shadow.

Place names on the Colorado Plateau are as colorful as the landscape: The Temple of Sinawava Wotan's Throne and Thor's Hammer; Elve's Chasm, Vasey's Paradise and Stillwater Canyon Island in the Sky, Robber's Roost and the Aquarius Plateau; Dirty Devil, Towers of the Virgin and Fern's Nipple; the Maze, Muley Twist and Furnace Flats.

Ultimately, all I can say is "come see this place" Get a sunburn, climb through an arch, discover a slot canyon, wade across a river, find a rattlesnake scare yourself. This is a land that will give your life back to you.

EARTH

I wish you were here! For this is a land that lies beyond the realm of description. It is a land where words, illustrations and science fail, a land that lies beyond the edge of myth.

This is a vast, hard-edged, bare bones landscape painted a thousand tints of watercolor-red. Cliffs, buttes, canyons, pinnacles, mesas, arroyos, spires and plateaus stretch to a rock-hard horizon. A horizon as clearly etched in this light as the crossbedded sandstone at your feet.

It is a land shaped almost exclusively by water, time and gravity, yet it is dry and timeless as a vacuum. Beneath a horizon oceanic in its flatness lies a twisted, eroded, upthrust, anticlined, block-faulted nightmare of geology. It appears as unchanging as infinity, yet everywhere is evidence of cataclysmic change. For these rocks tell tales of ancient lakes, streams, swamps, lagoons, seas and Saharan deserts. Mesas composed of oceanic limestones tower thousands of feet over frozen sand dunes and vice versa.

I like it!

WINTER STORM, CEDAR BREAKS N.M.

UPPER CATHEDRAL VALLEY, CAPITOL REEF N.P.

ASPEN GROVE, NORTH RIM, GRAND CANYON N.P.   12

MONUMENT CANYON, COLORADO N.M.

HOPI POINT, GRAND CANYON N.P.

LOMOKI RUIN, WUPATKI N. M.

KAIPAROWITS PLATEAU AND PADRE BAY, GLEN CANYON N.R.A.

FRESH SNOW, BRYCE CANYON N.P.

SUNSET, CEDAR BREAKS N.M.

MORNING, CORAL PINK SAND DUNES S.P.

PARUNUWEAP CANYON, ZION N.P.

STORM OVER CANYONLANDS N.P.

DELICATE ARCH, ARCHES N.P.

24

SUNRISE FROM SUNSET POINT, BRYCE CANYON N.P.

PAINTBRUSH, COLORADO N.M.

CHOCOLATE DROPS, THE MAZE, CANYONLANDS N.P.

VIEW FROM SUNSET POINT, CAPITOL REEF N.P.

VIEW FROM GREEN RIVER OVERLOOK, CANYONLANDS N.P.

ROCK

PEACH CANYON HOLLOW, ARIZONA

This is a land of rock, of rock upon rock. It is not some uniform monolithic granite landscape softened with a carpet of forest and meadow...oh no!

It is a sun-baked, wind-blasted landscape out of Dante's Inferno. Cracked and shattered and fractured into a million bizarre forms. Composed of shales and mudstones; painted deserts and trees turned to stone; sandstones and limestones and a million acres of petrified sand dunes.

To some it is hard as flint, ungiving, uncaring, broken and disheveled; to others it is soft and sensuous and gives freely of itself to those willing to receive its bounty. It is the God-awfulest place I have ever seen.

I like it!

WHITE HOUSE RUIN, CANYON de CHELLY N.M.

HOODOO, GOBLIN VALLEY S.P.

TURRET ARCH FROM NORTH WINDOW, ARCHES N.P.

PAINTBRUSH AND SANDSTONE, ZION N.P.

THE CORKSCREW, ANTELOPE CANYON

ROCK WALL, ESCALANTE CANYON

Here, in this region of bare rock, where water is little more than a rumor, everywhere there are the indications of its presence, of its creative hand in shaping and forming this arid land.

The distant rumble of summer thundershowers hints of change. Dehydrated sandy washes become raging red torrents of foaming liquid mud. Bone-dry cliffs become leaping waterfalls as water pursues a path of least resistence. Sun-scorched plateaus become turquoise mirrors as barely perceptible potholes fill with precious water, slowly percolating into the porous sandstone only to reappear years later at the seepline, thousands of feet below.

With water comes the rattling of cottonwood leaves, the songs of birds and the drone of insects. A musical tinkling gives evidence of a seep gently watering a lush garden of monkeyflower, columbine and maidenhair fern. Yet, just beyond the dripline lies the scorched earth of sand, prickly pear and yucca.

I like it!

RAINBOW OVER GRAYS PASTURE, CANYONLANDS N.P.

SLICKHORN CANYON, GLEN CANYON N.R.A.

PADRE BAY, GLEN CANYON N.R.A.

SPRING-FED WATERFALL, ZION N.P.

SAN JUAN RIVER AND MONUMENT VALLEY FROM MULEY POINT

THE NARROWS, ZION N.P.

BADGER CREEK RAPID, GRAND CANYON N.P.

SIPAPU BRIDGE, NATURAL BRIDGES N.M.

POTHOLE, GRANDVIEW POINT, CANYONLANDS N.P.

LAKE POWELL WATERS, GLEN CANYON N.R.A.

SUNRISE, TOROWEAP OVERLOOK, GRAND CANYON N.P.

SKY

RAINBOW AT SUNSET, GLEN CANYON NATIONAL RECREATION AREA

Often unnoticed in lesser locales, here in this red land the sky
becomes a physical presence, pressing down on the horizon, just
beyond reach. A deep cerulean blue sky that is as empty as space
one moment and host to monstrous boiling cumulo-nimbus the
next. Distance is telescoped; mountains and mesas float atop a
shimmering mirage of light.

Under the mad sun of mid-summer the sky seems almost cruel, a
great magnifying lens that intensifies the white-hot blaze of midday.
Seen at sunset in the winter of the year, it becomes a vast floating
watercolor awash with swirling color, alive.

Just when one suspects it of being a vast blue void, mocking and
jeering the hard-breathing hiker below, it suddenly brings the heart-
piercing song of the canyon wren or the good-natured laughing of
ravens flying in tandem with their own shadows along towering
golden walls of frozen sand.

I like it!

CROSSBEDDED SANDSTONE, ZION N.P.

SUNSET, PAINTED DESERT, PETRIFIED FOREST N.P.

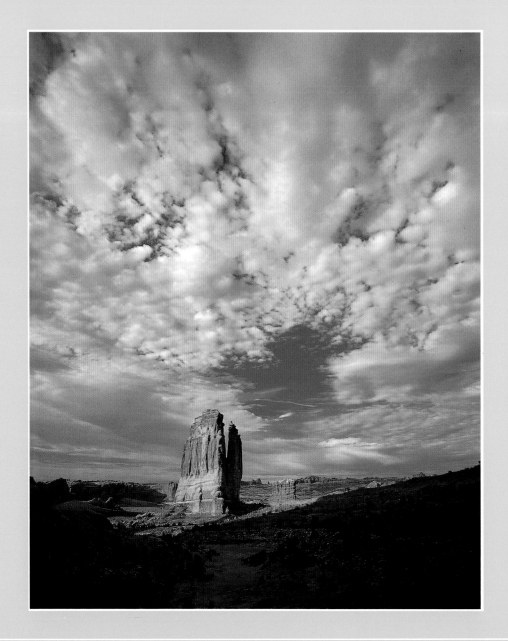

THE ORGAN, COURTHOUSE TOWERS, ARCHES N.P.

RAVEN, MONUMENT VALLEY T.P.

MOON AND CHECKERBOARD MESA, ZION N.P.

CLOUD AND CLIFF, CAPITOL REEF N.P.

MT. HAYDEN, POINT IMPERIAL, GRAND CANYON N.P.   70

PEEK-A-BOO ROCK, CAPITOL REEF N.P.

JUNIPER AT DEAD HORSE POINT S.P.

SUNSET, MATHER POINT, GRAND CANYON N.P.

VIEW FROM PANORAMA POINT, CAPITOL REEF N.P.

THOR'S HAMMER, BRYCE CANYON N.P.

MIDDLE EMERALD POOL, ZION N.P.

EVENING LIGHT, SAN RAFAEL SWELL

OWACHOMO BRIDGE, NATURAL BRIDGES N.M.

SUNRISE, ESCALANTE BASIN

RAINBOW AND CATHEDRAL ROCK, GLEN CANYON N.R.A.

CRESCENT MOON, QUEEN'S GARDEN, BRYCE CANYON N.P.

TSEGI OVERLOOK, AUTUMN, CANYON de CHELLY N.M.

ANGEL ARCH, CANYONLANDS N.P.

LANDSCAPE ARCH, ARCHES N.P.

SUNSET SILHOUETTES, GRAND CANYON N.P.

SUNRISE, MONUMENT VALLEY T.P.

I wish you were here! Here in this land of juniper, pinyon, aspen and spruce. Here in this land of sandstone, rattlesnake, raven, quicksand, badger, slot canyon, petrified wood, scorpion and Kokopeli. This land of scientific fact and shamanistic mystery.

I wish you were here, right here in this land of contradiction. This dry, sun-parched landscape carved and shaped by the force of raging water. This land of Wagnerian opulence and Zen-like simplicity.

This is a land which will wear you out, tear you down and leave you for the vultures. It will also fill your being, enrich your spirit and make your life worth fighting for. It is a helluva place to lose a cow and, God-almighty...I like it!

NOTES ON THE PLATEAU

by Jim Wilson

THE COLORADO PLATEAU

President Theodore Roosevelt once said in reference to the Grand Canyon, this is "one of the great sights which every American if he can travel at all should see". Grand Canyon is certainly deserving of such accolades, but his words could apply to the entire Colorado Plateau. This 130,000-square-mile geologic province teems with scenic treasures. Here water is both creator and sustainer of life. To be sure, man-made highways cut through this region, allowing easy access to admire Nature's handiwork, but the network of watercourses which tentacle the province represent the true method of transportation. This network is dominated by the Colorado River and its tributaries. On this high desert plateau water arrives infrequently. However, it is usually in quantities sufficient to spark flash flooding. It is this raw, unharnessed power which is responsible, in large part, for carving the chasms seen throughout this region. In addition to water the forces of cataclysmic uplift, grit-bearing winds, freeze-expansion-thaw and gravity have played contributing roles in sculpting the colorful ridges, mesas, buttes, pinnacles, bridges and arches of this land. Nearly two billion years of earth's history can be observed in its canyons. Stories of ancient oceans, lakes, deserts and lagoons are revealed in multi-hued sedimentary layers. Ages of vulcanism are described in igneous rocks and unanswered questions can be found in the metamorphic ones. Elevation plays a major role here, ranging from over 12,000 feet in the La Sal Mountains and San Francisco Peaks to little more than 1,200 feet at the point where the Colorado River emerges from Grand Canyon at Lake Mead. The biotic communities on the plateau are as diverse as these elevations would suggest.

HOODOO WITH CAP, GOBLIN VALLEY

HUMAN HISTORY

Although this area is one of the most sparsely populated regions of the country, it was home to numerous prehistoric Indian civilizations. Archaeological highlights include rock art, petroglyphs and pictographs found almost everywhere: pueblo ruins in Wupatki, Petrified Forest and Hovenweep, and incredible cliff dwellings in Canyon de Chelly, Navajo and Mesa Verde. These early Native Americans, as well as more contemporary settlers, had no part in the creation of this land but they most certainly learned to live on it. It is a harsh land in which the Anasazi and Fremont cultures were able to establish themselves, flourish for a period and then, for reasons not clearly understood, move on. All this occurred for approximately one thousand years prior to the fourteenth century. Some believe today's Hopi Indians and Pueblo peoples of Arizona and New Mexico descended from these civilizations.

With the westward expansion of the United States in the mid-1800's came settlers of European descent. Mormons were the most determined and successful pioneers of the Four Corners area. This region contains numerous historic sites that tell of their hardships, successes and failures. Of special note are Pipe Spring National Monument and the Fremont River area of Capitol Reef National Park. In both locations the National Park Service maintains the settler's fruit orchards and invites visitors to share in the harvest's bounty. In order to place this area's remoteness in proper perspective one must realize that in the continental United States this was the final region to be explored and mapped and includes the last mountain range named by Anglos, the Henry Mountains. Pages 17,25,37,83,91,93,94.

COLORADO NATIONAL MONUMENT

The Colorado River emerges from the hard stone of the Rockies and enters the Colorado Plateau at the northeast corner, near Grand Junction, Colorado. Here the tilted layers of the Uncompahgre Plateau have been eroded by flood waters into a series of stunning, steep-walled, box-shaped canyons which contain isolated rock monoliths rising from their floors. Colorado National Monument preserves this array of canyons carved into soft plateau rock. Westward views from Rim Rock Drive offer a view into the Grand Valley of the Colorado and across to the Book Cliffs. The river continues west, then southwest past the Uncompahgre and into Utah. Pages 13,29. On its north bank, near the La Sal Mountains and just before reaching Moab, is a land of natural arches.

ARCHES NATIONAL PARK

Arches National Park is home to the world's largest concentration of natural stone arches, numbering close to 2,000. The first question awed visitors ask usually pertains to how they were formed and why they occur in this geographic region. Answers to these questions are based on scientific speculation arrived at by careful study and research. The events which created Arches National Park, and all of the Colorado Plateau, occurred over a span of hundreds of millions of years, hence there are no eyewitness accounts or written records to substantiate circumstantial scientific evidence. This area lies atop a subsurface salt bed, thousands of feet thick in places, which was deposited across parts of the Colorado Plateau over 300 million years ago, the result of an ancient ocean which eventually evaporated.

NAVAJO PETROGLYPHS, CANYON de CHELLY

During succeeding millions of years sediments, up to a mile thick were deposited alternately by floods, winds and encroaching oceans. The underlying salt bed could not withstand the great downward pressure exerted by this sedimentation and it began to shift, buckle, liquify and subsequently, reposition itself. In some instances domes formed, sections dropped into cavities, or faults occurred. This subsurface movement shaped the Earth from below while erosion was removing sedimentary layers from above. As the earth was being reshaped by convoluting salt, fracture lines formed in this buried sandstone. Subsequent erosion removed exposed layers and enlarged sandstone cracks, isolating narrow fins. Alternate freezing and thawing resulted in

crumbling and flaking between fins; some even broke through. Cracking, gravity and weathering eventually widened holes to arch proportions (to be called an "arch" requires a minimum opening of three feet). In this area arches are formed in the orange-colored Entrada Sandstone layer. The same forces which gave birth to arches continue, bringing about their demise. As time passes they will collapse. One example is Landscape Arch. During the winter of 1941 a large block fell from its north buttress, enlarging its span from 291 feet to 306 feet. The next step in its demise occurred September 1, 1991, when a section of rock measuring 60 feet long by 8 feet wide by 4.5 feet thick broke away. The 4.5 foot loss reduced its thickness by more than 25%. How much longer can it withstand Nature's forces? In addition to natural stone arches, this park is home to many other stunning rock sculptures. Ancient dunes of the Navajo Sandstone layer can be seen "frozen" in place; eroded monoliths, balancing rocks and sandstone fins also vie for attention. All combine to make Arches National Park a unique portion of the Colorado Plateau. Pages 24,41,66,85,92. Continuing southwest some sixty river miles the Colorado River joins with the Green River in a land of canyons.

CANYONLANDS NATIONAL PARK

Canyonlands National Park preserves an immense wilderness of rock at the confluence of two great western rivers. Water, in all its forms, has been the primary sculptor of canyon country. Over millions of years it has carved layers of sedimentary rock into untold numbers of canyons, mesas, buttes, fins, arches and spires. At its center are two great chasms carved by the Colorado and Green Rivers. Surrounding the rivers are vast, yet quite different, areas of the park. Northward lies Island in the Sky, westward is the Maze, and to the east the Needles. Each possesses unique characteristics but share common origins, ecosystems and a wild spirit. Island in the Sky District is a high mesa, a platform from which to view the entire scene from horizon to horizon. Green River Overlook offers a view of the incredible work accomplished by waters originating in the

Wind River Range of Wyoming. Grandview Point, at the end of the road, allows a view of Monument Basin and the White Rim. Midway between Moab and Monticello is a paved road that leads to a Bureau of Land Management site which affords spectacular views of the Needles District. Seven miles further south U.S.Highway 191 meets Utah Highway 211, a route leading to Newspaper Rock, a Utah State Historic Monument preserving a significant panel of rock art. Continuing west on highway 211 brings one to the access point for the Needles District. Needles is home to a startling array of sculptured rock spires, arches, canyons, grabens and potholes. The Maze District, west of the Colorado River, is one of the wildest and most inaccessible areas in the United States. It is a perplexing jumble of canyons which has been described as a "thirty-square-mile puzzle in sandstone". Comparatively speaking, even the easily accessible areas of this 527 square-mile wilderness are seldom visited. Its interior regions are even less visited due to primitive four-wheel-drive roads and hiking trails leading into areas which do not have water sources. Due to its rugged remoteness, this park remains, for most travelers, an undiscovered gem. It is well worth devoting extra time to explore and extra care to protect and preserve this beautiful desert environment. Pages 23,25,28,30, 31,33,50,59,84. From the Needles District a short drive south brings you to an area of natural stone bridges.

SOUTH WINDOW, ARCHES N.P.

NATURAL BRIDGES NATIONAL MONUMENT

A natural bridge is carved by, and spans, a watercourse. The bridges of the Monument were carved into Cedar Mesa Sandstone, initially formed as windblown dunes. Over millions of years the watercourses that form White and Armstrong Canyons cut deep, looping gorges in the crossbedded stone. When a river forms a looping meander, almost circling back on itself, it creates a thin wall in which bridges can form. Floodwaters scrape away at both sides and percolating surface water works from above. Eventually the river breaks through and assumes the new, shorter path. As time passes, the river continues to enlarge the opening. Gravity, along with the freeze-expansion-thaw process, carves the underside and eventually the bridge crumbles. Natural Bridges contains three classic examples in three separate phases of development. Kachina is in its youthful stage, coarse and crudely sculpted, while Sipapu is in its mature years. Owachomo, no longer subjected to stream erosion, is experiencing old age. It is a thin, well sculpted formation, awaiting inevitable collapse from the force of gravity. Pages 58,79. Follow the water down canyon under the bridges and you will ultimately reach the heart of the Colorado Plateau.

GLEN CANYON NATIONAL RECREATION AREA

The "heart of the Colorado Plateau" designation makes perfect sense. By the time the Colorado reaches lower Glen Canyon, all other major tributaries, with the exception of the Paria, Little Colorado and Virgin Rivers, have merged with it. Here the Colorado River gathers its power from the far reaches of its drainage system. The culmination of this massive force was the carving of a magnificent slickrock canyon containing deep slots, tapestried walls, arches, alcoves and natural bridges. In 1956 the Bureau of Reclamation began the construction of the Glen Canyon Dam. Water began flooding Glen Canyon in 1963 and finally reached full pool level in 1980 submerging all river level features. The depths and walls of Glen Canyon now provide for the nearly 200 mile-long Lake Powell. The dam was born amid great controversy and compromise. It fulfills its goals of water storage and power generation while providing recreational opportunities. The resulting lake, with 1900 miles of shoreline, makes it possible to view those natural marvels and cultural features remaining above water line. Glen Canyon National Recreation Area was established in 1972 to administer and interpret the lake area and the nearly one million acres of surrounding desert and canyon country. Pages Cover, 14,16,18,51,53,60,62,81,95. Following Hall's Creek, a north shore tributary, brings you to a mind-boggling wrinkle in the earth's crust.

CAPITOL REEF NATIONAL PARK

Following the Notom-Bullfrog Road north from Bullfrog Marina you cannot miss a convoluted wrinkle rising high on the western horizon. This monocline, named Waterpocket Fold, continues northward forming a nearly impenetrable 100-mile north-south barrier. This massive buckle in the earth was created by the same incredible forces that built the Colorado Plateau some 65 million years ago. Capitol Reef National Park preserves the fold and its surrounding eroded jumble of cliffs, domes, spires, arches, monoliths and twisting canyons. The elongated southern section displays the fold in all its remote glory. Capitol Reef is a land of paradoxes. It is a dry land of uplifted rock and spectacular views while the sections that lie along Sulphur Creek and the Fremont River, parallel to Utah Highway 24, are lush riparian habitats which support a diverse ecological balance. The first human occupation dates back to 700 AD (the Fremont culture) and continues to the present time. Mormon settlers began farming the valley in the late 1800's, establishing productive fruit orchards. The National Park Service continues to preserve this culture and cultivate the orchards. The Cathedral Valley section of the park is one of the most remote and solitary places on the plateau. Golden eagles soar and stone monoliths tower over desert plains. There are few places on earth that can offer comparable peace and solitude. Pages 6,11,32,38,69, 71,74,88. Travel west to the high plateau country and discover the refreshing headwaters of the Paria River along with a fantasyland of red spires.

DOORWAYS, MESA VERDE N.P.

BRYCE CANYON NATIONAL PARK

High atop the Paunsaugunt Plateau, home to subalpine forests reminiscent of a high mountain range, lies the youngest sedimentary layer visible on the Colorado Plateau. This layer, the Claron Formation, was laid down approximately 50 to 60 million years ago as lime, mud and silt from shallow lakes and rivers. By the time this silty, impure limestone, long since transformed to stone, reached the surface through erosion, the pressure of uplift had left it fractured. Such fracturing renders soft limestones ripe for excavation. Averaging 8,000 feet above sea level, Bryce is subjected to heavy winter precipitation, usually half snow and half rain. Winter's freeze-expansion-thaw process combined with heavy summer thundershowers contribute greatly to Bryce's rapid erosion. Spread out before you is the magic of that erosion, forms of such intense colors and whimsical shapes that they inspired names such as Fairyland, Silent City, Thor's Hammer and Sinking Ship. Pages 19,27,75,82. If you could travel forty miles west while blindfolded or sleeping you would think you had never left Bryce Canyon.

CEDAR BREAKS NATIONAL MONUMENT

Consider for a moment that you could be fooled into thinking this was "West Bryce". It would not take long to learn otherwise. In spite of the similar geological formations the air is cooler and thinner. You are 2,000 feet higher, still atop the Claron Formation and still on the top step of the Grand Staircase. Similar forces created both Cedar Breaks and Bryce. Here at the highest reaches of the Colorado Plateau, in addition to hoodoos (small pinnacles or spires), one can find spruce-fir forests, montane meadows and, throughout the summer season, lush wildflower gardens. Traveling south from this northwest corner of the plateau you take the first step down the Grand Staircase. The Staircase is actually a series of cliffs, all descending from the north as the superimposed rock layers are eroded away. Some of these layers formed at the bottom of ancient seas, others on the coastal plains adjacent to the shifting ancient shorelines, still others as vast windblown sand dunes. The layers were uplifted during the last five to ten million years and are now being gradually eroded. Pages 9,20. Below Cedar Breaks are the headwaters of the Virgin River and the top of the next "step".

ZION NATIONAL PARK

High on the slopes of the Markagunt Plateau, the Virgin River begins its relentless flow to the Colorado River and Lake Mead. After slicing through the rim of the Kolob Terrace, the Virgin drops steeply through a series of sedimentary rocks and carves an incredible canyon...Zion. Where the Virgin River cuts through Navajo Sandstone, it forms steep, sheer cliffs and a "Narrows" up to two thousand feet deep and barely twenty feet wide. As the river cuts down through the Navajo and into the softer shales of the Kayenta Formation, the Narrows end and the portion of Zion Canyon most visitors see begins. Navajo Sandstone is made up of fine grains. They are loosely cemented and yield to erosion yet are rigid enough to support the deep gorge without collapsing. The underlying Kayenta Formation, however, easily washes away and subsequently undercuts the sandstone walls above. Eventually gravity prevails and the walls collapse, widening the canyon. This process is evident in the Temple of Sinawava at the Gateway to the Narrows. On the Colorado Plateau water is life. The Virgin River is a constant, year-round source of life-giving water, allowing the canyon to abound with a wide variety of flora and fauna. In addition to Zion Canyon, the park encompasses the finger canyons and terraces of the Kolob, the Great West Canyon of North Creek and the slickrock country surrounding its rims. Zion's high country is a study in sand, huge piles of sand grains loosely cemented into stone. 180 million years ago sand grains were piled in two-thousand-foot-deep dunes that once covered almost the entire Colorado

THE CASTLE, HOVENWEEP N.M.

Plateau. As winds changed direction in this ancient land, they blew the sand at angles to the previously deposited beds. Subsequently ancient seas encroached and the dunes became "frozen" in place, complete with the crossbedding pattern evident in such features as Checkerboard Mesa today. Pages 22,42,44,54,56,64,68,76. The next stop while descending the Grand Staircase is the Canyon of Canyons.

GRAND CANYON NATIONAL PARK

The Grand Canyon is an incredible spectacle, a classic example of erosion unequaled anywhere on Earth. The multi-hued slopes and cliffs of this chasm descend in a stunning panorama, culminating in the narrow Inner Canyon a mile below, where the Colorado River continues to cut ever deeper into the heart of the earth. It is more than a mere canyon. Its magnitude is equal to that of a major mountain range inverted into the earth. It is the grand climax of the Colorado Plateau, the sum total of all the power nature has released into this geologic province. Grand Canyon is a park made up of many ecological worlds, including the subalpine forests of North Rim, the ponderosa pine forests of South Rim, the high desert-like Tonto Plateau and a Sonoran Desert environment at river level. In addition, there are untold micro-climes interspersed throughout, each possessing unique characteristics. One can be overwhelmed by a first visit to Grand Canyon and feel comfortable after several visits, but few, if any, have ever gained a complete mastery of all that the park offers. Its allure and complexities demand that you return again and again in order to satisfy the thirst for understanding. Pages 12,15,57, 61,70,73,86.

PETRIFIED FOREST NATIONAL PARK

This park is renowned for two major features, a vast expanse of colorful landscape and the most complete fossil record of the Late Triassic terrestrial ecosystem found anywhere in the world. Recent paleontological research is piecing together an understanding of an ecosystem dating from 230 million years ago. No longer are the Park's abundant petrified logs seen merely as a beautiful collection of oddities. They are now known to be part of this ancient ecosystem, which represents an especially significant time in the evolution of life, a time of transition, when earlier forms were giving way to the earliest dinosaurs. The Park's most famous feature, its "forests of stone", are seen lying down and exposed by erosion. During the Late Triassic period this area was an expansive flood plain. Great trees were washed here from higher ground, then buried in silt, mud and volcanic ash. Water seeping through this bed gradually caused the replacement of wood tissue with silica. Following a more recent period of uplift, wind and erosion removed portions of the matrix, exposing thousands of the logs and countless fragments.

Facilities, now in the planning, will allow visitors to interact side by side with scientists conducting research into earth's evolution. Pages 26,65.

these gems deserve the same preservation we offer National Park lands. Pages 8,40 45,47,48,52,55,77,78,80.

STATE PARKS

While travelling through plateau country one should not overlook the many special places outside the National Park system. There are many fine state owned-properties, most notably in Utah. Several, such as Anasazi, Edge of the Cedars, Fremont Indian and Newspaper Rock, preserve significant prehistoric Indian cultural sites. Others offer visitors the opportunity to experience scenes of incredible beauty or geologic formations. Coral Pink Sand Dunes features a vast area of constantly changing wind-swept dunes. Dead Horse Point offers infinite views into and across canyon country. Goblin Valley is the home of contorted hoodoos eroded from the Entrada Sandstone layer. Here a vivid imagination can run wild. Kodachrome Basin is a land of red spires and chimneys reaching into the sky. Pages 21,39,72,90.

BUREAU OF LAND MANAGEMENT (BLM)

One must keep in mind that special places of the Colorado Plateau are not confined to those set aside as designated parks. Surrounding most of these parks are other federally owned lands, many of them equally exciting. The task of managing a large percentage of this stunning scenery is given to the Bureau of Land Management. For the most part, visitors to the Southwest know what to expect of National Parks and Monuments. Books and videos depict the famous scenery as well as offering tips on how to enjoy them. But for the adventurous, the true joy of discovery comes when one happens upon an unexpected gem. For every National Park on the Colorado Plateau there are untold numbers of stunning locations on BLM properties, places like Fisher Towers, the Escalante area, Grand Gulch, Paria Plateau, Calf Creek, Factory Butte, Muley Point, Canyon Rims, San Juan River and the San Rafael Swell. These less visited places are often remote, consequently many well-seasoned plateau visitors prefer them. Keep in mind that

PRIMROSE AND SAND

NATIONAL FORESTS

In addition to BLM locations there are also federal lands designated as National Forests. In most cases, they are the high-elevation forest lands that border, or are islands in, the vast high deserts of the Colorado Plateau. The northwestern boundary is formed by Fishlake National Forest. Its many lakes feature fine fishing for several varieties of trout, including Mackinaws weighing up to 35 pounds. On the west can be found Dixie National Forest. It features unusual rock formations and extensive aspen groves. Autumn is an especially enjoyable season in Dixie. Kaibab National Forest encompasses three sides of Grand Canyon National Park. It is a high country of pine, spruce and aspen and is home to a diverse population of wildlife including the rare Kaibab squirrel, elk and mule deer. To the south is Coconino National Forest, containing the volcanic San Francisco Peaks and an ecosystem that ranges from desert to tundra. The eastern side of the plateau is home to Manti-La Sal National Forest. Significant features are the peaks of the La Sal and Abajo Mountains. These, in addition to the Henry Mountains and Navajo Mountain, are classified by geologists as laccoliths, volcanos that 'fizzled'. They represent igneous and metamorphic formations that could not muster enough energy to penetrate the weight of thousands of feet of sedimentation. Erosion, during the successive ten million years, has stripped away those overlying layers exposing these harder, more resistant 'islands'. The La Sal and Abajo Mountains offer summer visitors a cool respite in dense forests of aspen, pine, fir and spruce. Page 10.

SUGGESTED READING

Abbey, Edward; Desert Solitare; Peregrine Smith, Inc.; 1981.
Abbey, Edward; Slickrock; Sierra Club Books; 1971.
Barnes, F.A.; Utah Canyon Country; Utah Geographic Series; 1986.
Crampton, Gregory C.; Standing Up Country; Alfred A. Knopf; 1964.
MePhee, John; Encounters With the Archdruid; Farrar, Straus and Giroux; 1971.
Porter, Eliot; The Place No One Knew; Sierra Club Books; 1963
Stegner, Wallace; Beyond the Hundredth Meridian; University of Nebraska; 1982.
Telford, John; Coyote's Canyon; Peregrine Smith, Inc.; 1989.
Trimble, Stephen; The Bright Edge; Museum of Northern Arizona Press; 1979.
Zwinger, Ann; Run River, Run; University of Arizona Press; 1984.
Zwinger, Ann; Wind in the Rock; Harper & Rowe; 1978.

Natural History Associations, non-profit organizations chartered by Congress to aid the National Park Service, are excellent sources of published material which interpret the Park or Monument in which they operate. Their publications may be found at the sales areas in visitor centers or by contacting them directly.
Bryce Canyon Natural History Assoc.- Bryce Canyon, Ut. 84714.
Canyonlands Natural History Assoc.- 125 West 200 South, Moab, Ut. 84532. (Arches, Natural Bridges, BLM, USFS)
Capitol Reef Natural History Assoc.- Torrey, Ut. 84775.
Colorado National Monument Assoc.- Fruita, Co. 81521.
Glen Canyon Natural History Assoc.- P.O. Box 581, Page, Az. 86040.
Grand Canyon Natural History Assoc.-P.O. Box 399, Grand Canyon, Az. 86023.
Petrified Forest Museum Assoc.- P.O. Box 277, Petrified Forest, Az. 86028.
Zion Natural History Assoc.- Zion National Park, Springdale, Ut. 84767.

PHOTOGRAPHIC CREDITS

Mary Allen: 22,47,78.
Robert Barr: 31,85.
Charles Cramer: 34,45,46.
Eduardo Fuss: 86.
George H.H. Huey: 53,83.
Gary Ladd: Cover,14,18,48,52,62,73,81.
Mark and Jennifer Miller: 67.
William Neill: 24,36.
Jeff Nicholas: 4,11,16,20,21,28,33,38,39,54,56,59,64, 66,69,71,72,75,79,90,92,93,94,95.
Pat O'Hara: 30,65.
Larry Pierce: 15.
Greg Probst: 87.
Randall K. Roberts: 82.
Norm Shrewsbury: 10.
Dan Sniffin: 43.
John Stevens: 42.
John Telford: 12.
Tom Till: 8,9,17,19,23,25(2),29,51,60,61,84.
Larry Ulrich: 13,26,55,57,80.
Jim Wilson: 6,27,32,37,40,41,44,50,58,68,70, 74,76,77,88,91.
Copyrights to all photographs remain with the artist.

CREDITS

Text by Jeff Nicholas
Introduction by Lynn Wilson
Preface by Jeff Nicholas
Staircase Notes by Jim Wilson
Graphic Design by Jeff Nicholas
Map by Jeff Nicholas
Edited by Ardeth Huntington

Layout and graphic design performed on a Macintosh® SE utilizing Aldus PageMaker® and Microsoft Word®. All texts set in Palatino and Optima Typefaces by MacinType, Fresno, Ca. Color separations and printing coordinated by Interprint, Petaluma, Ca.